SUSTAINABILITY CONSIDERATIONS
FOR PROCUREMENT TOOLS AND CAPABILITIES

SUMMARY OF A WORKSHOP

Dominic A. Brose and Lynette I. Millett, *Rapporteurs*

Committee on Fostering Sustainability Considerations into
Public and Private Sector Procurement Tools and Capabilities

Science and Technology for Sustainability Program
Policy and Global Affairs

Computer Science and Telecommunications Board
Division on Engineering and Physical Sciences

NATIONAL RESEARCH COUNCIL
OF THE NATIONAL ACADEMIES

THE NATIONAL ACADEMIES PRESS
Washington, D.C.
www.nap.edu

THE NATIONAL ACADEMIES PRESS   500 Fifth Street, NW   Washington, DC 20001

NOTICE: The project that is the subject of this report was approved by the Governing Board of the National Research Council, whose members are drawn from the councils of the National Academy of Sciences, the National Academy of Engineering, and the Institute of Medicine. The members of the committee responsible for the report were chosen for their special competences and with regard for appropriate balance.

This summary report and the workshop on which it was based were supported by Contract No. xx001-xx972 between the National Academy of Sciences and the U.S. General Services Administration. Any opinions, findings, conclusions, or recommendations expressed in this publication are those of the author(s) and do not necessarily reflect the views of the agency that provided support for the project.

International Standard Book Number-13:   978-0-309-26259-0
International Standard Book Number-10:   0-309-26259-3

Additional copies of this report are available from the National Academies Press, 500 Fifth Street, NW, Keck 360, Washington, DC 20001; (800) 624-6242 or (202) 334-3313; *http://www.nap.edu*.

Copyright 2012 by the National Academy of Sciences. All rights reserved.

Printed in the United States of America

THE NATIONAL ACADEMIES
Advisers to the Nation on Science, Engineering, and Medicine

The **National Academy of Sciences** is a private, nonprofit, self-perpetuating society of distinguished scholars engaged in scientific and engineering research, dedicated to the furtherance of science and technology and to their use for the general welfare. Upon the authority of the charter granted to it by the Congress in 1863, the Academy has a mandate that requires it to advise the federal government on scientific and technical matters. Dr. Ralph J. Cicerone is president of the National Academy of Sciences.

The **National Academy of Engineering** was established in 1964, under the charter of the National Academy of Sciences, as a parallel organization of outstanding engineers. It is autonomous in its administration and in the selection of its members, sharing with the National Academy of Sciences the responsibility for advising the federal government. The National Academy of Engineering also sponsors engineering programs aimed at meeting national needs, encourages education and research, and recognizes the superior achievements of engineers. Dr. Charles M. Vest is president of the National Academy of Engineering.

The **Institute of Medicine** was established in 1970 by the National Academy of Sciences to secure the services of eminent members of appropriate professions in the examination of policy matters pertaining to the health of the public. The Institute acts under the responsibility given to the National Academy of Sciences by its congressional charter to be an adviser to the federal government and, upon its own initiative, to identify issues of medical care, research, and education. Dr. Harvey V. Fineberg is president of the Institute of Medicine.

The **National Research Council** was organized by the National Academy of Sciences in 1916 to associate the broad community of science and technology with the Academy's purposes of furthering knowledge and advising the federal government. Functioning in accordance with general policies determined by the Academy, the Council has become the principal operating agency of both the National Academy of Sciences and the National Academy of Engineering in providing services to the government, the public, and the scientific and engineering communities. The Council is administered jointly by both Academies and the Institute of Medicine. Dr. Ralph J. Cicerone and Dr. Charles M. Vest are chair and vice chair, respectively, of the National Research Council.

www.national-academies.org

COMMITTEE ON FOSTERING SUSTAINABILITY CONSIDERATIONS INTO PUBLIC AND PRIVATE SECTOR PROCUREMENT TOOLS AND CAPABILITIES

Dara O'Rourke (Chair), Assistant Professor, University of California, Berkeley
Peter Graf, Chief Sustainability Officer, SAP
Patrick Mallet, Founder, Credibility Director, ISEAL Alliance
H. Scott Matthews, Research Professor, Carnegie Mellon University
Teresa Neal, Senior Purchasing Manager, Church & Dwight, Co., Inc.

Staff

Marina Moses, Director, Science and Technology for Sustainability Program
Jon Eisenberg, Director, Computer Science and Telecommunications Board
Dominic Brose, Program Officer, Science and Technology for Sustainability Program
Lynette I. Millett, Associate Director, Computer Science and Telecommunications Board
Jennifer Saunders, Program Officer, Science and Technology for Sustainability Program
Dylan Richmond, Research Assistant, Science and Technology for Sustainability Program
Pat Koshel, Senior Program Officer, Science and Technology for Sustainability Program
Emi Kameyama, Senior Program Assistant, Science and Technology for Sustainability Program

Preface and Acknowledgments

In December 2011 the National Research Council's Science and Technology for Sustainability Program, in collaboration with the Computer Science and Telecommunications Board, held a workshop organized by the Committee on Fostering Sustainability Considerations into Public and Private Sector Procurement Tools and Capabilities. As sustainable purchasing becomes increasingly important in both the public and private sectors, tools that can facilitate the sustainable acquisition of goods and services will play a critical role in meeting sustainability objectives. Participants at the workshop (listed in Appendix B) assessed the current landscape of sustainable procurement tools, identified opportunities and emerging needs for enhanced or new tools, identified potential barriers to progress (such as interoperability), and explored potential solutions. Participants also considered the procurement workforce and associated training required to realize the full benefits of these tools. Workshop participants included users of procurement tools (including federal, state, and local government and industry), experts in sustainable procurement, developers and users of various types of data, and individuals from companies that develop and provide procurement tools and related software.

This document was prepared by the workshop rapporteurs as a factual summary of what occurred at the workshop. The examples and topics discussed in this report are limited to what was discussed at the workshop. For instance, although there are many examples of sustainable purchasing efforts in private industry and the public sector, the report describes only those that were explicitly discussed at the workshop. The statements made in this volume are those of the rapporteurs and do not necessarily represent positions of the workshop participants as a whole, the steering committee, the Science and Technology for Sustainability program, or the National Academies. This workshop summary is the result of substantial effort and collaboration among several organizations and individuals. We wish to extend a sincere thanks to each member of the steering committee for their contributions in scoping, developing, and carrying out this project. The project would not have been possible without financial support from the General Services Administration (GSA).

This report has been reviewed in draft form by individuals chosen for their diverse perspectives and technical expertise, in accordance with procedures approved by the National Academies' Report Review Committee. The purpose of this independent review is to provide candid and critical comments that will assist the institution in making its published report as sound as possible and to ensure that the report meets institutional standards for quality and objectivity. The review comments and draft manuscript remain confidential to protect the integrity of the process.

We wish to thank the following individuals for their review of this report: Leonardo Bonanni, Sourcemap; Scot Case, UL Environment; Wendy Gordon, Natural Resources Defense Council; and Verena Radulovic, U.S. Environmental Protection Agency.

Although the reviewers listed above have provided many constructive comments and suggestions, they were not asked to endorse the content of the report, nor did they see the final draft before its release. Responsibility for the final content of this report rests entirely with the rapporteurs and the institution.

Dominic A. Brose
Lynette I. Millett
Rapporteurs

Contents

1	OVERVIEW	1
2	GOVERNMENT EFFORTS	5
3	SOURCING AND MATERIALS	13
4	TOOLS AND TECHNOLOGY FOR SUSTAINABLE PURCHASING	21
5	WORKFORCE AND CULTURE	29

APPENDIXES

A	WORKSHOP AGENDA	33
B	REGISTERED PARTICIPANTS LIST	39
C	BIOGRAPHIES	43
D	EXAMPLES OF FEDERAL AGENCY PROCUREMENT SYSTEMS AND GREEN PURCHASING SYSTEMS	55
E	THE FEDERAL LIFE-CYCLE ASSESSMENT (LCA) DIGITAL COMMONS	61

1

Overview

Federal laws, regulations, and executive orders have imposed requirements for federal agencies to move toward the sustainable acquisition of goods and services, including the incorporation of sustainable purchasing into federal agency decision making. In particular, two Executive Orders—EO 13423, *Strengthening Federal Environmental, Energy, and Transportation Management*, signed in 2007; and EO 13514, *Federal Leadership in Environmental, Energy, and Economic Performance*, signed in 2009—include specific goals and objectives for sustainable purchasing by agencies. Federal government green purchasing efforts, however, can be traced back to at least EO 12759 signed by President Bush in 1991. The federal government spends tens of billions of dollars on goods and services each year. Since the federal government is such a significant player in the market, its move to incorporate sustainable procurement practices could have a profound impact on the types of products being developed for the market as a whole.

The General Services Administration (GSA) has played a key role in furthering sustainable procurement practices throughout the federal government. GSA is responsible for formulating and maintaining government-wide policies covering a variety of administrative actions, including those related to procurement and management. GSA has several ongoing activities related to sustainable procurement, many of these related to Section 13 of EO 13514, which directed the agency, in coordination with other key agencies, to assess the feasibility of working with the federal supplier community—vendors and contractors that serve federal agencies to measure and reduce greenhouse gas emissions in the supply chain while encouraging sustainable operations among suppliers. GSA has also been actively developing programs to assist federal agencies in making sustainable procurement decisions. As federal agencies cannot directly fund the development of sustainable procurement tools, they are particularly interested in understanding how to foster innovation and provide incentives for collaboration between developers and users of tools for sustainable purchasing throughout the supply chain. The training of procurement professionals is also a priority for these agencies.

Agencies also face challenges related to whether and how suppliers collect and provide data on the sustainability of their operations. For example, suppliers may not be collecting data on their greenhouse gas emissions or may not be willing to provide the data to agencies due to concerns about confidentiality or competition. Agencies are actively evaluating opportunities to encourage suppliers to disclose relevant data on sustainability performance. Ultimately, procurement professionals may need to access these types of data to make decisions about sustainable acquisition activities for their agencies.

To assist efforts to build sustainability considerations into the procurement process, the National Research Council appointed an ad hoc committee to organize a two-day workshop that explored ways to better incorporate sustainability considerations into procurement tools and capabilities across the public and private sectors. The workshop was designed to help participants assess the current landscape of green purchasing tools, identify emerging needs for enhanced or new tools and opportunities to develop them, identify potential barriers to progress, and explore potential solutions. Participants also considered the workforce and associated training required to realize the full benefits of these tools. Participants at the workshop included: users of sustainable procurement tools (including federal, state, and local governments and industry), experts in sustainable procurement, developers and users familiar with open data, and individuals from companies that develop and provide procurement tools and software. The workshop provided an opportunity for participants to discuss challenges related to sustainable purchasing and to developing new procurement tools.

Presenters discussed tools currently used in sustainable procurement, such as databases for ecolabels and standards, codes, or regulations; calculators that track environmental footprints; software for traceability of materials; and life-cycle assessment (LCA) software. Some participants viewed the development of apps for smartphones and tablets as a useful emerging capability with significant potential for incorporating procurement tools and applications. Other nontechnological tools were discussed as well, such as procurement policies, frameworks, rating systems, and materials or product indexes. In considering existing tools and requirements for new ones, several overarching themes and associated criteria emerged from the workshop presentations, breakout groups, and discussion sessions, including:

- Integration of sustainability criteria for products and services
- Data management and cloud computing
- System integration and interoperability
- Encompassing the full extent of the procurement process

Participants discussed the enormous amount of data that would be required to give procurement professionals access to real-time information in order to make up-to-date, effective decisions. Integrating procurement systems with other systems—especially financial ones—was discussed by many participants as key to new tools for sustainable procurement. Many participants also noted that agreement on a standard language—semantics and syntax—is important in furthering progress in systems integration and ultimately to entire sustainable purchasing networks.

In addition, some participants pointed out that culture and workforce training are critical to ensuring the success of any new tools developed for sustainable purchasing systems. Sustainable procurement results from a complex system of suppliers, vendors, program managers, contracting officers, and procurement professionals. Some participants noted that making information and tools available at points earlier in the procurement process—not just at the purchasing phase—would allow sustainable procurement to be approached more holistically. Empowering procurement professionals to make more informed decisions was also suggested as key to making change in these areas. Pilot projects, training, and collaboration were presented as ways to build "buy-in" from procurement professionals and leadership, which is important in ensuring that sustainable purchasing practices and tools are used to their full potential.

2

Government Efforts

Many of the sustainable procurement activities within the federal government have been spurred by the 2009 Executive Order EO 13514, *Federal Leadership in Environmental, Energy, and Economic Performance*, which requires federal agencies to develop sustainability goals that focus on making improvements related to environmental, energy, and economic performance. As part of this effort, the General Services Administration (GSA) is working to integrate sustainability into its purchasing decisions. The Section 13 Interagency Working Group, created under Section 13 of EO 13514, is evaluating the feasibility of working with the federal vendor and contractor community to provide information to assist agencies in tracking and reducing greenhouse gas emissions (GHG) related to the supply of products and services to the government.

According to workshop presenter Stephen Leeds[1] from the Office of the Administrator at GSA, the agency purchases about $95 billion in goods and services annually, including 12 million products through 18,000 vendors, making the agency well positioned to influence the federal government's purchasing decisions. GSA's goal is to have a supply chain that is sustainable throughout. Sustainable procurement, Mr. Leeds added, is about "making smart investments in products that provide better services" for a longer period of time. Achieving a sustainable supply chain requires an understanding of the environmental "hotspots" within an industry—in other words, the components of the supply chain with

---

[1] Senior Counselor to the Administrator at the time of the workshop.

the largest environmental impact. The agency's thinking about sustainable purchasing is evolving, with a focus on life-cycle approaches, return on investment, risk mitigation, intentionality, and partnership, he said.

GSA is adopting a life-cycle approach to sustainable purchasing as it "bridges the silos of disposal and acquisition," Mr. Leeds explained. The agency is sending a clear signal to the private sector that the focus is broader than the individual environmental impacts of purchasing decisions. Partnering with the private sector, as well as with state and local governments, is important to the agency's work and its goal of achieving a sustainable supply chain.

Nancy Gillis from GSA's Federal Supply Chain Emissions Program Management Office (PMO) described how GSA and other agencies have worked to advance sustainable acquisition in the federal supply chain. GSA does not view the concept of "sustainability" as synonymous with the "environment," Ms. Gillis said, but considers it a broader issue that encompasses economic and social issues as well. The agency is approaching procurement decisions by prioritizing products' life-cycle return on investment and by considering environmental, economic, and social benefits and costs. In other words, the agency is trying to balance the need to reduce energy use, resource use, and environmental impacts while also taking into account economic considerations.

Ms. Gillis discussed the office's other activities, including collaborating with industry and supporting and managing the Sustainability in Procurement Fellowship Program. The program introduces fellows to the concept of sustainability and provides an overview of the federal government's activities around sustainable procurement. Another effort is the GreenGov Supply Chain partnership, which was designed to increase the energy efficiency of vendors and contractors' supply chains and to reduce their GHG emissions. The partnership resulted from a GSA report that found sustainability considerations, especially GHG emissions data, should be used in the federal procurement process, and that agencies should engage the vendor community to track and reduce GHG emissions through a collaborative, transparent, and deliberative process.[2]

The Section 13 Interagency Working Group is currently evaluating and recommending ways to advance sustainable acquisition throughout the federal government, Ms. Gillis explained. For example, after evaluating whether it is feasible for the federal contractor community to provide GHG emissions data related to the supply of products for use in government procurement decisions, the working group recommended that suppliers not be required to provide complete inventories of their GHG

---

[2] General Services Administration (GSA). 2010. Executive Order 13514 Section 13: Recommendations for Vendor and Contractor Emissions. Washington, DC.

emissions. Instead, the group recommended that the government, as an incentive, inform suppliers that GHG emissions data could be considered by agencies when making procurement decisions.

Alison Kinn Bennett from the Environmental Protection Agency's (EPA) Office of Pollution Prevention and Toxics described the activities of a subgroup of the Section 13 working group that focused on product standards and ecolabels. This subgroup, which involves GSA, EPA, Department of Defense (DOD), National Institutes of Health (NIH), United States Department of Agriculture (USDA), and other agencies is responsible for ensuring that the product-related acquisition goals of EO 13514 are met by providing guidelines for selecting environmental sustainability standards or ecolabeling programs. The subgroup views standards setting as a pyramid, as shown in Figure 2-1. Environmental and health data and tools are the base, with standards and incentives for green products built on those, Ms. Kinn Bennet said. From that, a system to verify standards is established, so that ultimately buyers are able to more easily find green products with effective, reliable standards.

Ms. Kinn Bennett noted that the subgroup's work has had several phases, including grounding; developing draft guidelines; assessing the guidelines using a survey of standards and consultation among federal agencies; holding listening sessions with stakeholders; and preparing a report. The assessment phase of the subgroup's work focused on existing U.S. and international protocols for standard setting and environmentally preferable product claims and verification methodologies. A survey conducted by the subgroup identified about 80 guidelines for selecting standards in ecolabels. Those guidelines were then categorized into five general areas:

- Standard setting: how the standard was created
- Standard substance: the content, relevance and effectiveness of standard criteria
- Conformity assessment: how the standard was created and whether it was third-party verified or compliance was self-declared
- Program management: how the program is managed, governed and operated
- Market penetration: the extent to which it is used and recognized in the market

Ms. Kinn Bennett noted that going forward, it will be important for agencies to address trade-offs and assess environmental impacts across media and life-cycle stages when making sustainable acquisition decisions. The subgroup wants to encourage more holistic, comprehensive thinking on standards and criteria development, she said.

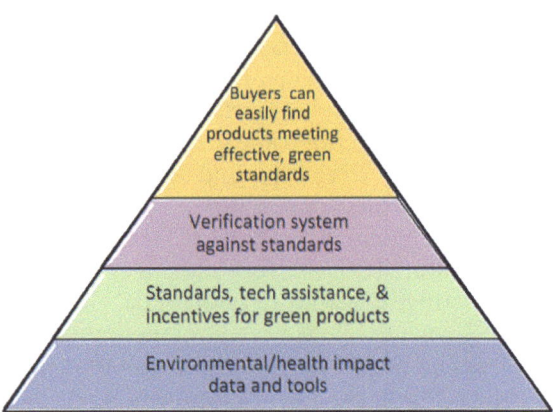

FIGURE 2-1 Development of effective, green standards for products that buyers can readily identify.
SOURCE: Alison Kinn Bennet presentation, December 7, 2011.

Josh Silverman from the Office of Sustainability Support at the Department of Energy (DOE) and Karen Moran from the Defense Logistics Agency (DLA) with the Department of Defense described sustainable procurement efforts at their respective agencies. According to Dr. Silverman, DOE relies heavily on its contractors to implement and integrate sustainability into procurement decisions; the agency establishes requirements, but contractors are responsible for implementing them. DOE also relies heavily on effective communication and information exchange; for example, the agency has an ongoing working group with hundreds of active members who regularly exchange information on best practices. The agency also tracks data on the use and procurement of green products and is trying to further integrate sustainability requirements into its contracting, with the goal of using sustainable procurement as a strategy to drive improved mission performance. The agency now requires contractors—particularly those providing construction and custodial services—to review contract actions for ways to integrate sustainability requirements.

Additionally, DOE conducted a quasi-"hotspot" analysis to assess both the products in highest demand and available standards, Dr. Silverman said. The Green Buy Program within DOE offers awards to contractors and vendors that purchase these priority products. He added that DOE is trying to incentivize and reward behavior that integrates sustainable approaches. It will be necessary to better quantify the benefits of these activities so that their impact can be understood and used to make the business case for sustainable procurement, he said; a purely compliance-oriented approach may not be as effective in moving this effort forward.

Karen Moran described the role of the DLA as a combat logistics support agency and noted that much of the agency's work is contracted. DLA is continually trying to incorporate sustainability into its procurement decisions and those of its contractors, Ms. Moran said. She described several ongoing sustainable procurement programs at the agency, including efforts to encourage procurement of bio-based products; a program that returns unused medications to vendors; and a variety of alternative fuel initiatives, including efforts to increase use of renewable energy and biodiesel. Another initiative is incorporating environmental information into an electronic purchasing system known as the Federal Catalog System. This information includes environmental attribute codes that indicate a sustainable characteristic for a specified product.

Ms. Moran described a recent joint GSA/DOD sustainable procurement project that kicked off with a meeting on November 17, 2011. The meeting brought together representatives from both agencies to discuss translating policy into action, the need to integrate sustainable procurement into the course of business, life-cycle assessment, and the value of synchronizing GSA and DOD to enhance standardization. Participants at the meeting discussed the need for compelling messaging that would cast sustainable procurement in terms of an agency's mission, Ms. Moran said. Discussion at the November meeting also explored potential ways to motivate behavioral change related to sustainable procurement along with the recognition that training alone cannot create culture change. Finally, Ms. Moran noted that there was discussion of the need for a toolbox, which should be scaled in complexity and size for different types of procurement, as well as the need for a mechanism to test products and provide feedback.

Edward Rau from the Division of Environmental Protection at the National Institutes of Health (NIH) spoke about efforts to incorporate sustainable procurement at NIH. His department directly purchases, funds, or influences purchases related to health care, food, drugs, and biomedical research; however, NIH has few sustainable criteria for these products and no effective tools to help make such purchases more sustainable, he noted. The information on sustainable purchasing that is available consists primarily of static reference documents distributed across many separate Web sites, which typically cannot be applied by procurement professionals to perform, facilitate, or track purchasing transactions. One area where NIH would benefit from a sustainable procurement tool would be in purchasing replacement freezer units that are 10 percent more energy efficient than the current units on NIH's Bethesda campus, Captain Rau said. This would save NIH about $1 million a year in electricity costs. According to

> **Elements of a Sustainable Purchasing Tool:**
>
> ✓ Provide an authoritative, one-stop sustainable shopping reference
> ✓ Centralized and automated
> ✓ Interactive
> ✓ Simple, two-step search and buy function
> ✓ Have data collection and reporting functions
> ✓ Link directly to other procurement systems
> ✓ Characterize transactions as compliant or noncompliant to FAR
> ✓ Track desirable, sustainable attributes
> ✓ Link to agency approvals and published literature
>
> Edward Rau, National Institutes of Health, December 8, 2011.

a recent survey, the current freezer units account for 29 percent of total electricity use, $12 million a year in costs, and releases of about 59,000 metric tons of carbon dioxide equivalence per year.

Another gap in sustainable procurement tools is in identifying hazardous toxic or polluting substances, Captain Rau noted. From a public health perspective toxicity is one of the most important criteria in sustainable acquisition, he said. Although the interim Federal Acquisition Regulations (FAR) require procurement of products and services that are nontoxic, this has been difficult to implement, due in part to a lack of clear definitions and toxicity data. To address gaps in available information, Captain Rau has spearheaded the development of a "substance of concern" list that would restrict or prohibit the government from procuring products or services that contain or release listed substances of concern. The proposed list could be used as an interim screening and selection method until better data and methods of comparative toxicology are developed. The list would inventory substances by their chemical abstract service registry number to reduce synonym confusion, be derived from other well-established listings such as EPA's Integrated Risk Information System (IRIS)[3] or the Consumer Product Safety Commission's Safer Products lists,[4] and characterize the listed substance as banned or restricted in certain uses. Where available, it would also list alternatives for the substance of concern.

Another area with gaps that a procurement tool could address is the tracking and reduction of greenhouse gas emissions in the supply chain,

---

[3] *www.epa.gov/iris*

[4] *www.saferproducts.gov*

added Captain Rau. Metrics, emissions accounting requirements, and boundary definitions would all need to be established for such a tool. Full life-cycle data would also be needed so that the focus is not just on the embodied greenhouse gas of the product but also its full life cycle. He gave the example of sulfur hexafluoride, which appears on the substance of concern list, not because it is toxic per se but because it is a potent greenhouse gas.

Another major gap in sustainable procurement efforts is consideration of a product's end of life. Currently, procurement tends to focus on a product's recycled material content or bio-based materials rather than on its reusability, recyclability, or biodegradability. These considerations are critical for meeting net zero goals that are becoming more widely adopted, Captain Rau said.[5]

He described his view of the elements that would be included in an effective sustainability-oriented procurement tool. It would provide an authoritative, one-stop sustainable shopping reference for all products and services, be very centralized, and be automated. It would be interactive and not merely a reference companion or compilation. It would have a simple, two-step search-and-buy function; purchasers would search for a specific product or service and then be directed to where to buy a compliant product. This would save time by eliminating the need to search for applicable requirements and products that conform to them, as is currently done. The tool would eliminate the need for purchasers to understand complex, rapidly-changing, sustainable procurement requirements, thus minimizing training needs. It would also have data collection and reporting functions and would link directly to other procurement systems to avoid multiple data-entry errors. In addition, it would characterize a transaction as compliant or noncompliant with any regulations, such as the Federal Acquisition Regulations (FAR), and track how well a product or service adheres to other desirable but not mandatory sustainable attributes. It is critical that users have confidence in the quality of the data, and be assured that purchases made with the tool meet all applicable requirements, Captain Rau stressed. Lastly, for scientific and medical applications, it would be critical that purchases of medical supplies and devices directly link to agency approvals, such as Food & Drug Administration (FDA) approvals, and to published medical literature.

---

[5] The DOD has a primary goal focused toward net zero, meaning net zero energy use, water use, and waste. The aim is to start with reduction, then progress through repurposing, recycling, energy recovery, and lastly disposal. Other agencies are moving toward this concept of net zero. See: army-energy.hqda.pentagon.mil/netzero/.

3

Sourcing and Materials

Procurement is often thought of as simply the act of purchasing goods and services by processing a requisition, receiving an invoice, and then making a payment; however, many workshop participants emphasized that procurement should be thought of as a much broader process, from generating and soliciting requirements all the way to closing out a contract (Figure 3-1). Participants also noted that procurement tools to help foster sustainable purchasing could be integrated into all aspects of the process. Such integration could allow for better accounting of the actual costs of goods and services acquired through the procurement process. That process, however, is often considered as starting with the procurement professional. Nancy Gillis from GSA's Federal Supply Chain Emissions Program Management Office explained that many steps in the procurement process occur prior to the actual purchasing phase, but typically it is only at the point of purchasing where the conversation about sustainability begins.

As Jonathan Rifkin from the District of Columbia's Office of Contracting and procurement noted, procurement professionals ensure that hundreds of legal, regulatory, and policy requirements are implemented in an open and public way. Unless these professionals are given a clear idea of how to help meet sustainability goals when purchasing products, the process can essentially turn into a check-the-box exercise, making implementation of any sustainable strategy difficult to do in a meaningful way.

By the time a procurement professional first sees a solicitation, Mr. Rifkin noted, it is already too late in the process to affect any significant

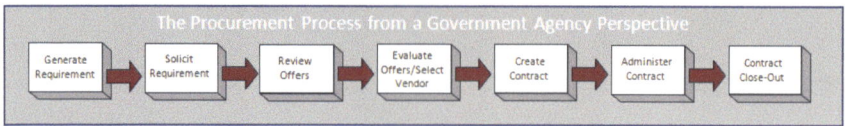

FIGURE 3-1 The procurement process at a federal agency.
SOURCE: Nancy Gillis presentation, December 7, 2011.

sustainability-related changes to products they purchase. It would be more influential to insert sustainable specifications earlier than the purchasing phase— for instance, when the statements of work for a particular procurement are being developed. New procurement tools could help, Mr. Rifkin said. A workshop participant also suggested that procurement professionals could be evaluated in part based on their performance in meeting certain sustainability criteria. Mr. Rifkin added that it would be possible to assess performance based on sustainability criteria only as long as clear specifications, priorities, and procedures are provided.

Stephen Gordon from Old Dominion University echoed the theme that earlier stages in the procurement process are important and that sustainability considerations could be incorporated at any point along the supply chain and at any stage in the life of a product, including disposal. Life-cycle assessments (LCAs) aim to assess the environmental, economic, and social impacts of a product throughout its lifespan. Attributing costs to different stages in the life cycle of a product is one use of LCAs, although even if a more sustainable product will save money over time, upfront costs can be a barrier to purchasing it. The cost of disposing of products, especially those containing hazardous components, is seldom considered during purchase, Dr. Gordon noted. If procurement professionals incorporate this consideration, they may decide to purchase a product that is costlier up front, but which may ultimately result in savings over the entire life cycle. Some participants discussed the possibility of financing upfront costs and then using cost savings to pay that amount back—an example of a revolving funding scenario that could help address high initial costs.

Dr. Gordon also noted that it is important to account for sustainability historically by looking at a product's past performance to quantify savings and process improvements, referred to as the return on investment. Forecasting savings is important so that contracting officers are not penalized up front for purchases that may initially be more expensive but that will save on externalities in the future. There are qualitative case studies but few peer-reviewed scientific data on these forecasting methodologies, which would benefit from further analysis and quantification, Dr. Gordon said.

There is opportunity for more work on life-cycle assessments, as Dara O'Rourke from the University of California, Berkeley, pointed out. LCAs have primarily been used by academics and consultants and have been less useful for decision makers within procurement agencies or companies, he said. However, some modern LCA tools and approaches offer better visibility into the entire supply chain. A challenge is bridging the gap between the academic black box and general use so that information from a product's anticipated life cycle is used in LCA tools in a way that helps decision makers make more informed choices about a product, Dr. O'Rourke noted. Participants offered potential solutions to make information from LCAs and LCA tools more accessible, such as providing such tools at little or no cost to the public sector.

Some participants also noted that the choice of technology platform and availability of data can affect how widely a tool is used. For instance, modern Web applications and associated mobile apps could help move technology and tools forward. In addition, a shift toward more accessible or "open" data could enable an ecosystem to develop around agreed-upon data structures and standards. Participants noted that conventional systems made up of proprietary data and code may not be the dominant model much longer. As one participant noted, such shifts are currently exemplified by USDA's National Agriculture Library's Federal Life-Cycle Assessment Digital Commons, which is coordinating the collection of LCA data and information on organization, management, dissemination, and preservation in a way that makes data available to all users (Appendix E).

SUPPLY CHAINS

Edan Dionne from IBM Corporation described the company's Supply Chain Social Responsibility Audit Program. Third-party auditors are used to audit suppliers against IBM's Supply Chain Code of Conduct, which addresses labor, safety, environment, ethics, and social responsibility. Suppliers found to be noncompliant are required to implement improvement plans, she said; IBM will discontinue using suppliers that fail to improve after implementing such a plan. Ms. Dionne noted the responsibility IBM feels to develop its supply base by working to educate suppliers and asking them to focus on environmental and social issues and impacts.

It is especially helpful to work with first-tier suppliers, clarifying expectations by providing specific tasks for improvement, Ms. Dionne added. For example, IBM recently imposed a management system requirement that allows suppliers to examine environmental and social impacts more holistically and requires them to implement programs to address noncompliance in these areas. Many suppliers want to do the right thing,

Ms. Dionne noted, but they need to have the organization, policies, and capability in place to sustain their performance over time.

Patrick Mallet from ISEAL Alliance noted that a big challenge is integrating data and ensuring consistency, both in how data are understood and also in the language used to convey meaning. One tool that has been developed to address such challenges is the Nike Considered Index. Lorri Vogel from Nike Inc. stated that the company focuses on designers because, as the earliest group in the product-development process, designers have the largest impact downstream. As such, they need design tools and systems that allow them to make better choices. Peter Graf from SAP also noted that the connection between the design phase and recycling is key, and that in terms of procurement tools, there are overarching elements that apply regardless of the sector or type of products purchased (Figure 3-2).

Nike's Considered Index focuses on materials, waste, solvents, and innovation, said Ms. Vogel. She gave as an example a particular Nike shoe, which scored well on environmentally preferred materials and reduced weight but not as well on solvents. The key issue was bonding dissimilar materials, such as the plastic plates in the shoe. The design team came up with a new way to bond the plates to the midsole, improving the shoe's score on solvents and also scoring well on innovation.

FIGURE 3-2 The connection between the design phase and recycling is key; there is opportunity for information technology and software to take a model that is based on linear processes and make it cyclical.
SOURCE: Peter Graf presentation, December 8, 2011.

The Nike Material Sustainability Index (MSI) is another tool used by Nike to organize information on the materials they use. Nike's Jim Goddard noted that the company tracks materials from their sources—whether the material was grown in a field or extracted from an oil well—to the product factory. This takes into account all of the early life-cycle stages of the material, which can account for up to an estimated 60 percent of the environmental impact of a product. The idea behind the MSI, Mr. Goddard said, is to ensure that useful information is available at the right time in the product design process, so that designers and developers can make decisions about the best material while creating a product. Trade-offs are necessary when making decisions about what material to use, and that information needs to be integrated as a normal part of the design process.

A ROLE FOR ECOLABELS

Some participants noted that different types and amounts of information are needed at different points in the procurement process. For example, the procurement professional may not need to know all of the details about the sustainability of a particular supplier's supply chain, as long as someone within the purchasing organization with a broader perspective incorporates that information. Ecolabels, standards, and certifications are tools that can convey some kinds of sustainability-related information. Ecolabels could also address the desire for a single, simple notation that represents an agreed-upon optimization of the environmental, social, and economic attributes of a product or service, some participants said.

Being able to represent these products' attributes in a simple, easily understood label could enhance the efficiency of the purchasing process. Certifications could be especially useful in areas where "green-washing[1]" take place. Alicia Culver from the Responsible Purchasing Network emphasized that purchasers who are not trained in sustainability could easily fall for what looks like a sustainable product based on claims from a label. Claims such as "natural" or "earth friendly" are commonly used on consumer products, Ms. Culver noted. She also explained that claims on many retailer-created ecolabels are inconsistent, which increases confusion. Harmonization and agreement on the claims and criteria would help clarify matters.

Anastasia O'Rourke from Big Room Inc. said that her company had tracked the growth of ecolabels back to 1954 when the first such label—a "safe" toy label in Germany—emerged. Since then, the number and

---

[1] Practice of making an unsubstantiated or misleading claim about the environmental benefits of a product, service or technology.

complexity of ecolabels has increased significantly; labels can be multi-attribute or single-attribute based, life-cycle based, specific to a region, or global. Ms. Kinn Bennet from Environmentally Preferable Purchasing at EPA described ecolabels or standards as residing at the tip of a triangle, which would be all the information some people, such as procurement professionals, may need (see Figure 2-1). But information that goes beyond the label—such as information from LCAs and standards-setting and verification procedures—can also help, she said. Single-attribute and multiple-attribute certifications are currently being used, with single-attribute labels focusing on just one area, such as human health or environmental impact. Choosing one attribute often requires a trade-off with another attribute, Ms. Culver noted. Multi-attribute certifications can address several areas, such as life-cycle impacts, human and environmental toxicity, packaging, and performance.

An ecolabel could serve as a visualization that might be enough for the purchaser, explained Dara O'Rourke. With current technology, however, a deeper dive into more data and information could be readily available, though not necessarily required each time a purchaser selects that product, he said. Josh Saunders from Greencurement commented that certifications exist for some products for which a lot of environmental or health information is available; however, for many other products such data do not exist, and so neither do the certifications. Not many ongoing efforts in the public or private sector are attempting to solve this problem, he added. Also, some consumer efforts are trying to increase transparency in the market, creating an opportunity to demand better data to inform purchasing decisions, Mr. Saunders said. Gathering more information, standardizing data, and increasing interoperability of systems are all potential components of this approach.

Mr. Saunders also commented that one way to move forward would be to convince a number of large institutional, influential purchasers to gather better information that can be used by all to make more informed decisions. Dara O'Rourke noted that a broad framework is needed to encompass all of this information from products, standards, and labels; purchasers need information from suppliers, who need information from their suppliers, and so on down the supply chain. It is key that this information is meaningful, relevant, and accurate, he said. Other participants noted that such a framework should be developed in an iterative fashion; as data are collected and used, it would likely become clear some data are not useful and that different information is needed. Thus, it would be important to feed this information back into the process and to adapt it accordingly. As one participant expressed, we have to build systems that allow us to learn over time and ensure that the right information is sought out.

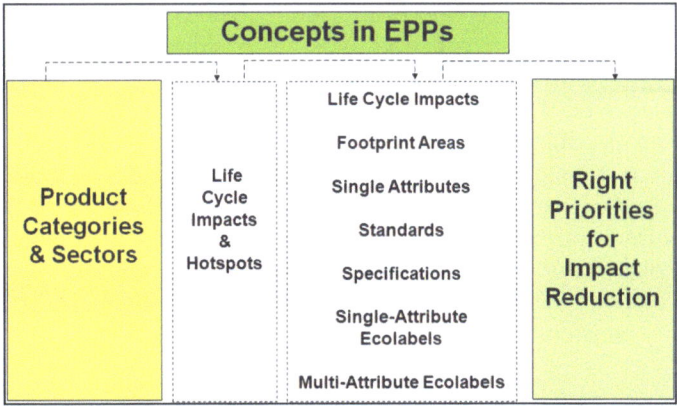

FIGURE 3-3 Terms and concepts in environmentally preferable purchasing (EPP) that feed into a framework for sustainable products.
SOURCE: Yalmaz Siddiqui presentation, December 8, 2011.

Yalmaz Siddiqui from Office Depot discussed an example of a framework for sustainable products he has worked on with the Green Products Roundtable, a voluntary stakeholder group working to reduce confusions over the "green" marketplace and improve the production and buying decisions of manufacturers, institutional purchasers, and consumers.[2] The framework attempts to incorporate many concepts surrounding the idea of a sustainable product, such as LCA outcomes, attribute-based concepts, standards, and single- and multi-attribute ecolabels. Work needs to be done on definitions so that everyone has the same understanding of the concepts being discussed, he added. Those concepts and definitions then need to be incorporated into procurement tools. The first step in the group's approach to developing the framework was to identify the terms and concepts used in major environmentally preferable purchasing policies (EPPs). This involved identifying the attributes most looked for and the ecolabels most frequently used (Figure 3-3). The next step was to determine how to prioritize the different ways of reducing the impact of that product, where "impact" might refer to water use, greenhouse gas emissions, or some other environmental impact.

Mr. Siddiqui summarized the main ideas in the framework—an LCA that identifies types of life-cycle impacts, such as abiotic depletion or

---

[2] Keystone Center. 2012. The Green Products Roundtable. [online.] [Available at: http://www.keystone.org/spp/environment/Green-Products-Roundtable] [Accessed April 24, 2012].

toxicity, and four broad "footprint" areas that focus on materials and waste, water use, energy use, and toxicity. The framework then connects attributes that purport to reduce certain types of life-cycle impacts. Mr. Siddiqui stressed that data will be needed in order to understand what sectors and product categories are driving the highest life-cycle impact types. The framework itself is a tool of sorts, he added, but it could also serve as a reference for developing standards. In addition, it could function as a foundation for a procurement tool to help clarify which attributes, ecolabels, or behaviors reduce impacts in meaningful ways. The framework could also provide a way to distinguish between criteria used in making decisions.

4

Tools and Technology for Sustainable Purchasing

Different types of tools are used in procurement, whether in the private sector or in the federal system. These tools may be simple, such as specific ecolabels (described in Chapter 3), or more complex, such as tracking and reporting software or customized search engines used for purchasing and maintaining inventories. Complicating matters is that there are many different attributes of products, a variety of stakeholders with diverging opinions as to what attributes are important, trade-offs between attributes, a range of methodologies and terminology to define them, and many different technical platforms to convey this information (Figure 4-1).

Edward Rau from the Division of Environmental Protection at the National Institutes of Health (NIH) said that NIH's success in implementing sustainable procurement will depend on the availability of clear, complete criteria and on data that enable suppliers to access sustainable procurement criteria and distinguish compliant products from others. Success will also depend on purchasers selecting the most sustainable and competitive products in efficiently placed procurement transactions, and on procurement managers collecting and consolidating the acquisition data needed to assess how well the requirements have been implemented. Chris O'Brien from American University described several important factors that should be considered as new tools or technology for procurement are developed: stakeholder engagement, comparative capabilities, ability to integrate with other systems, alignment with policies, and format and aesthetics.

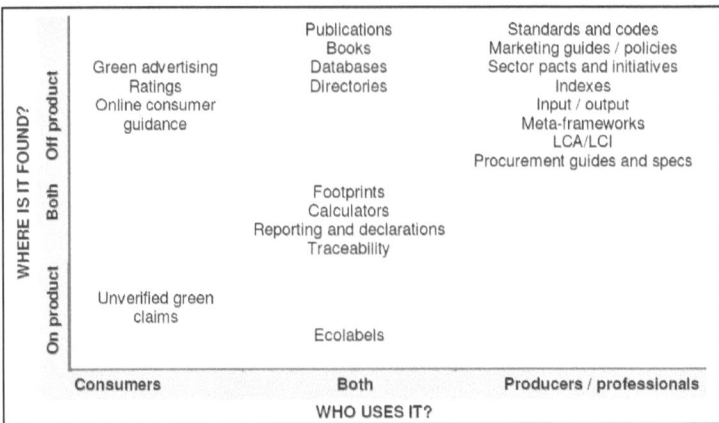

FIGURE 4-1 The landscape of product sustainability tools.
SOURCE: Anastasia O'Rouke presentation, December 7, 2011.

Three programs familiar to Mr. O'Brien in a university setting are the Sustainability, Tracking, Assessment, and Rating System (STARS),[1] Leadership in Energy and Environmental Design (LEED),[2] and American College and University Presidents' Climate Commitment (ACUPCC).[3] All three programs are voluntary, Mr. O'Brien commented, and have successfully become de facto standards for universities—a result he attributed to the time invested in engaging stakeholders as the programs were developed. That investment resulted in immediate "buy-in" so that when the programs were released, they were readily taken up by stakeholders. Similarly, the openness of a tool—whether it is public or private in design—is important, he said; for example, the information provided to STARS and ACUPCC are public and available online, but data for LEED are generally not. A program is also informative if it allows benchmarking against peers, Mr. O'Brien added; with STARS or ACUPCC users are rated or provided a score and are able to compare themselves against peer scores in that system.

Mr. O'Brien emphasized that a tool used for procurement needs to be able to be integrated with other systems and especially with financial tools. Ideally, financial decision making and sustainable procurement should be part of the same tool, so that comparisons of data are made within a single integrated system. Also commenting on integration was

---

[1] *stars.aashe.org*

[2] *www.usgbc.org*

[3] *www.presidentsclimatecommitment.org*

> **Better Sustainable Purchasing Tools:**
>
> ✓ Integrated with financial systems
> ✓ Integrated with other sustainability-related systems
> ✓ Openness - online and optionally public
> ✓ Aesthetically beautiful to motivate users
> ✓ Enables behavior change
>
> Chris O'Brien, American University, December 8, 2011.

Josh Saunders from Greencurement, who said that having to use multiple systems for a purchase is a challenge to sustainable procurement. Right now, he noted, one might have to go to supplier Web sites, a certification Web site, internal spreadsheets or an internal Web site, a manufacturer's Web site, and of course, one's own procurement systems to gather the necessary information for a single sustainable purchase. Integration is needed so that new tools can share data and information in an accessible and usable fashion, Mr. Saunders said.

Peter Graf of SAP noted that as soon as people have to use systems running at different companies, the process slows down dramatically. These systems should work more effectively with each other, he said. Another important aspect of integration is that procurement is not a self-contained process; it is connected to many components of the organization, such as sourcing, contracting, financing, analytics, and auditing. Procurement tools that incorporate all parts of the process and integrate information on the financial impact would be very useful.

None of the three tools described earlier—STARS, LEED, or ACUPCC—have attained this financial integration, said Mr. O'Brien; however, they have all incorporated policies that require sustainable purchasing, even though they cover different aspects of it. In developing policy language to enable sustainable procurement of paper, for example, Mr. O'Brien said that in some cases he has to compare and contrast three different sets of guidance.

Usability of tools is also important, he added, noting that STARS is Web-based and straightforward to view and navigate. LEED certifications and ACUPCC, by contrast, are based on spreadsheets, and managing the voluminous amount of data and numerous tabs can make data gathering for these programs challenging. Engaging people with these tools is important, he said; to achieve sustainability goals, people need tools they can relate to and use readily or behavior will not change very quickly.

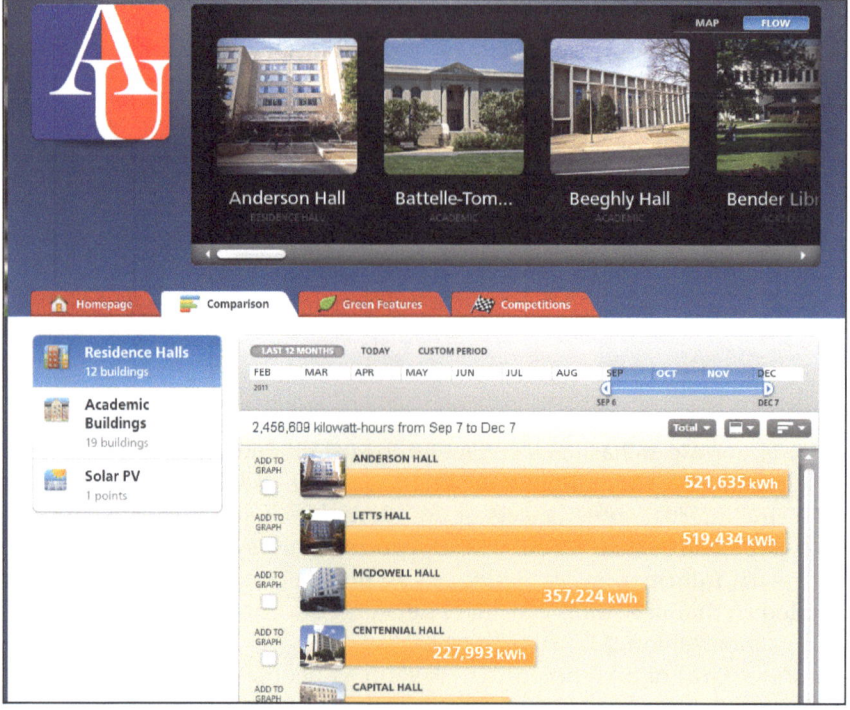

FIGURE 4-2 Screenshot of energy efficiency dashboard used at American University.
SOURCE: Chris O'Brien presentation, December 8, 2011.

Well-designed dashboards are another way to represent and foster integration, Mr. O'Brien added, presenting an energy dashboard that incorporates energy purchasing decisions into a graphical display. These data are projected on screens in every building on American University's campus, and occasionally energy efficiency competitions are held between buildings (Figure 4-2).

EMERGING REQUIREMENTS AND CRITERIA

Tools that are designed to work within or take into account an entire procurement process, not just the purchasing step, will be critical (see Figure 3-1), said Dr. Graf, who commented on emerging functional requirements, or, as he stated, "the evolution of procurement." Emerging functional requirements are focused on compliance, active management of suppliers, and optimization. Even more important is the recognition that procurement is not simply one company or organization procuring products from suppliers; rather, it is an entire economy encompassing a network of companies and organizations.

Addressing compliance entails looking at, among other things, the bill of material, which helps to ensure that supplier codes of conduct for regulatory compliance are met, Dr. Graf said. Addressing compliance also often includes aggregating data across an entire network of suppliers. One active management approach is to use a questionnaire or other check-the-box type format, such as that done to generate an Energy Star rating. For example, a company can take the bill of material of what was purchased and, based on data and information related to the material, convert energy use into a carbon footprint.

The optimization stage is a significant challenge, Dr. Graf noted. It requires understanding the complete supply chain for a product, the most relevant attributes, and how those attributes can be compared with those of products provided by another supply chain. Optimization is where comparisons and benchmarks become important. The benchmarks are sometimes between products and other times between suppliers. Dr. Graf commented that the transition from compliance to active management to optimization used to be considered in the context of a single company. It has since expanded to the whole supply chain of that company, and now to entire networks of companies and organizations and their respective supply chains.

Josh Saunders from Greencurement noted that all stages of the procurement process—defining purchasing requirements, contract RFP writing, deciding what to buy, contract administration, compliance, reporting, and measurement—are ripe for tool development.

Tools could be developed to overcome the current barriers to sustainable purchasing, such as information overload, Mr. Saunders said. For example, although ecolabels are important for sustainable purchasing, there are hundreds of ecolabels in the market. Purchasers already need to understand a great deal of information, and new information is constantly being generated. Tools that provide the right information to the right users at the right times would greatly assist procurement professionals, he said.

Related to information overload is the need for procurement professionals to make trade-offs among the different criteria and policies guiding them, Mr. Saunders added. Procurement professionals are primarily focused on optimizing price, availability, and the performance or quality of a given product. A tool is needed to allow sustainable criteria to be integrated into this focus so that procurement professionals can find products that are competitive with or lower in price than non-green products and that perform as well or better.

> **Procurement Components with Opportunity for New Tool Development:**
>
> ✓ Purchasing requirements
> ✓ Request for proposals (RFPs)
> ✓ Purchase decisionmaking, e.g., trade offs
> ✓ Contract administration
> ✓ Compliance with sustainability policies
> ✓ Measurement and tracking
> ✓ Reporting
> ✓ Overcoming barriers to sustainability: information overload, cost assessment, and establishing criteria
>
> *Josh Saunders, Greencurement, December 8, 2011.*

Integration and interoperability are also important, and shared semantics and application programming interfaces need to be developed where appropriate. Interoperability between tools is about translation between tools, said Anastasia O'Rourke of Big Room; it is fundamentally about how what one person says is translated for another person to understand. The potential benefit of making this work is increased data flow and accessibility of information. Another benefit would be increased accountability, because effective interoperability could allow information to be tracked across different systems. It could also allow for an increase in the scale and scope of accessible information and allow for better comparisons, helping purchasers gain more clarity. To move this forward, she stated, purchasers need to identify necessary tools and types of format; develop common glossaries and classification systems or methods for translating between systems; collaborate to map and compare systems; develop data frameworks and reporting templates; develop the ability to export data as appropriate to other systems; and participate in various multi-stakeholder forums. With so many stakeholders interested in sustainability issues, she added, purchasers need to be part of those conversations.

Procurement tools will increasingly incorporate more data into their functions, Ms. O'Rourke said. If, for instance, a tool displays a green button to indicate a good product, the data that helps determine whether that green button should be displayed may be only available elsewhere, or the metric indicated by the button may depend on near-real-time information. Such situations may require access to other systems and real-time connectivity.

Dr. Graf also commented on cloud computing and mobility. More

TOOLS AND TECHNOLOGY FOR SUSTAINABLE PURCHASING 27

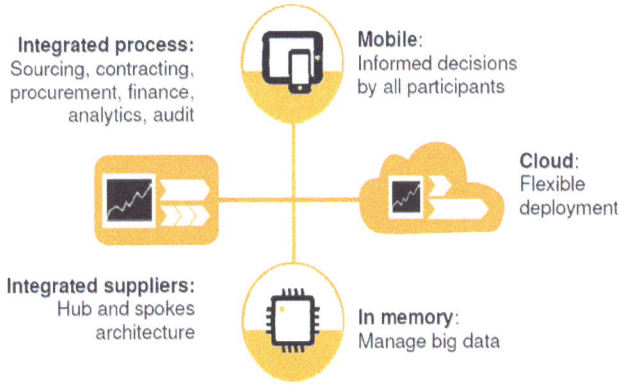

FIGURE 4-3 Technical requirements for sustainable procurement.
SOURCE: Peter Graf presentation, December 8, 2011.

and more, he noted, companies want to leverage cloud-based tools and systems in part because they can usually be deployed more quickly. Also, individuals want information to be accessible wherever they happen to be, not only through their desktop computers but from mobile devices. A procurement manager who needs to approve a specific purchase, for example, could see from glancing at his or her mobile device that a requisition is urgent. Figure 4-3 summarizes what many participants described as the technical requirements for sustainable procurement, including data management.

COMMON THEMES GOING FORWARD

During the last session of the workshop, participants discussed major gaps in knowledge for sustainable procurement, positive or negative attributes that characterize sustainable procurement, and what those attributes should be moving forward. The following are themes and issues many participants observed to be relevant and in need of further attention as sustainable procurement continues to be adopted by private and public sector organizations.

- Semantics and units of measurement were raised as an issue, especially regarding relevance and materiality. For example, even if the same unit of measurement is being used for the rating of a given product, the rating and information presented within a single product category are often not reported in the same away across

organizations. Increasing cohesiveness in the measurement and language used, such as by using product category rules, would bring more clarity to ratings and reports on products and be useful to purchasers.
- Certifiers could be urged to disclose more information and data from the companies they are certifying so that the process is more transparent, many participants felt.
- Many participants observed a disconnect between certifications/ecolabels and outcomes. The actual environmental health or societal impacts of products bearing such labels are typically not tracked and accounted for in any systematic way. A related issue is how inputs relate to the output of certified products, and how products could be designed with more consideration of outcomes.
- Harmonizing definitions on product types or moving toward unique identifiers like a sku number would allow different purchasers and procurement systems to readily identify the same product. This could also improve interoperability between technologies and data sharing among systems.
- In addition to the purchasing phase of procurement, many participants felt that contracting is also important. However, it may not be possible to address outcomes in this context; it may be better to address sustainability concerns by writing constraints into a statement of work. In this case, tracking and evaluating work performance in an effective way would be a more accurate assessment of sustainability practices.

5

Workforce and Culture

Behind the ecolabels, standards, tools, and technology are the procurement professionals and other staff who make the decisions to move sustainable procurement programs forward. Alicia Culver from the Responsible Purchasing Network noted that one best practice organizations can use is to institutionalize sustainable procurement, which ensures top-level support. Procurement professionals do not want to make decisions for end users unless the policy is clear to all stakeholders, Ms. Culver noted. Organizations need to determine a process for collaboration among different parts of the organization, adopt a policy with clear goals and reporting requirements, and establish communication and outreach strategies. Collecting baseline data at the beginning of a program is important for demonstrating the success of a program, showing improvement over time, and identifying cost impacts and savings, all of which could help demonstrate the case for a responsible purchasing program.

Also, Ms. Culver commented, there seems to be a movement toward decentralization of purchasing, which can pose challenges to sustainable procurement. When the decision-making process is centralized around green and sustainable products and services, an organization can often negotiate better prices. It is harder to aggregate demand when each agency or department is ordering on its own. Also, when purchasing is centralized, it is easier to control and monitor purchases, which simplifies the education and training process.

Participants discussed needs for training in the workforce, and one

> **Best Practices for Instilling Sustainable Procurement in an Organization and Workforce:**
>
> ✓ Determine a process for collaboration within an organization
> ✓ Adopt a policy with clear goals and reporting requirements
> ✓ Establish communication and outreach strategies
> ✓ Track information and establish a baseline at the beginning of a program
> ✓ Centralize procurement activities in the organization
>
> Alicia Culver, Sustainable Purchasing Network, December 7, 2011.

participant noted that the roles and responsibilities for different workers need to be clearly identified so that the proper level of training can be targeted to that group for best results. Another aspect of this, said another participant, is that a culture change is needed in addition to training. One solution is to have a pilot-scale project in order to foster buy-in from some, and then expand the effort to all, a participant suggested. Once a small example is out there, another participant added, it could lead to larger efforts and eventually to buy-in from leadership.

Training should be presented and viewed as education, and it should be a two-way activity, commented one participant. Receiving feedback, both data-driven and qualitative, helps point out opportunities for improvement and reveals whether the training had the right focus. One participant urged that culture change, especially empowering procurement professionals to make more informed choices and provide guidance, should be incorporated into the strategic vision of organizations; instilling sustainability principles will not be successful unless it is connected to the vision of the organization. Also important are promoting proper practices and giving recognition to the people who are doing things the right way, some participants noted.

Jonathan Rifkin from the District of Columbia's Office of Contracting and Procurement presented the case for an interagency, jurisdiction-wide team capable of addressing issues around sustainable procurement. The team should include a procurement professional who can move the process forward and who understands how to write contract language that can be readily applied. An environmental expert will also be needed, such as a representative from the jurisdiction's environmental unit. Such expertise would help inform the purchaser, who may not be able to make environmental value judgments for that particular jurisdiction. Mr. Rifkin also said that people from budget and finance would be needed on the team to overcome the notion that sustainable products are more expensive. Also important are people who could help recognize when a product with higher initial costs would result in savings over time. Such savings, he noted, could possibly fund this type of program in a way that it becomes a self-funding mechanism. In Mr. Rifkin's view, another important member of team would be someone from the office of the mayor (or city administrator or governor), to make clear to everyone that leadership supports the new approach.

Finally, having someone from the supplier community can be beneficial. This allows local businesses to be apprised of sustainable initiatives and any new requirements when it is time for a bid for particular types of products, Mr. Rifkin added; such information helps them prepare and be ready with any needed specifications. Pulling together this team ultimately gains an element of buy-in from this very politically powerful group of people, and turns into allies a group of individuals who before may have resisted sustainability initiatives. Such an effort would have many benefits, such as streamlining the sustainable purchasing process by giving procurement professionals clear and precise guidance about the process and making them a partner in it. This type of program could also make tracking purchases easier, Mr. Rifkin said; effectively tracking purchases is difficult, and without tracking it is hard to measure progress. Such programs would also allow jurisdictions to be creative, he said; states and local communities are the laboratories of the country, and giving them the tools and infrastructure to make decisions could allow for them to come up with distinctive solutions that will help everybody. Mr. Rifkin concluded that at the end of the day, bringing this group to the table and asking them to speak to these issues builds buy-in, which is probably the most difficult thing to achieve because of different priorities, needs, and other pressures.

Appendix A

Workshop Agenda

Fostering Sustainability Considerations into Public and Private Sector Procurement Tools and Capabilities

A National Academies Workshop

December 7-8, 2011
20 F St. Conference Center
20 F St. NW
Washington, D.C.
Conference Room A

Wednesday, December 7, 2011

8:30 AM	Welcome and Goals of the Workshop
	Marina Moses, Director, Science and Technology for Sustainability Program

	Dara O'Rourke, Chair, Workshop Planning Committee, University of California, Berkeley

Session I: Procurement Processes—Challenges and Opportunities

Moderator: Dara O'Rourke, Chair, Workshop Planning Committee, University of California, Berkeley

8:45 AM	GSA Comments on Sustainable Procurement
	Steven J. Kempf, Commissioner, Federal Acquisition Service, General Services Administration

9:00 AM	Section 13 Interagency Working Group Overview: Sustainable Procurement and Processes
	Nancy Gillis, Director, Federal Supply Chain Emissions Program Management Office General Services Administration

9:15 AM   Best Practices in Sustainable Procurement
             Alicia Culver, Director, Responsible Purchasing Network

9:45 AM   Section 13 Product Standards and Ecolabels Sub-working Group
             Alison Kinn Bennett, Senior Advisor on Product Sustainability, Environmental Protection Agency

             Brennan Conaway, Contracting Officer, General Services Administration

10:15 AM   BREAK

10:45 AM   Breakout Groups

             What are purchasers' challenges and needs across the lifecycle of purchasing? What are some examples of hitting barriers when trying to buy "green"? What would be the design criteria in developing the ideal tool?

             Moderator: Leo Bonanni, Chief Executive Officer, Sourcemap

             Location: Conference Room A

             What information about products is needed regarding sustainable attributes, whether directly and through certification and ecolabels?

             Moderator: Alison Kinn Bennett, Senior Advisor on Product Sustainability, Environmental Protection Agency

             Brennan Conaway, Contracting Officer, General Services Administration

             Location: Board Room

11:45 PM   Breakout Group Report Back to Plenary
             Moderator: Dara O'Rourke, Chair, Workshop Planning Committee, University of California, Berkeley

             Location: Conference Room A

12:30 PM   LUNCH

APPENDIX A

Session II: Fostering Sustainability in Procurement Processes
Moderator: Teresa Neal, Senior Purchasing Manager, Church & Dwight, Co., Inc.

1:30 PM	Sustainability in Procurement Processes—Data Needs and Challenges *Anastasia O'Rourke, Co-Founder, Big Room*
2:00 PM	Fostering Sustainability into the Procurement Process *Stephen Gordon, Professor, Old Dominion University*
2:30 PM	BREAK
2:45 PM	Federal Perspectives on Sustainable Procurement *Josh Silverman, Director, Office of Sustainability Support, Department of Energy* *Karen Moran, Environmental Protection Specialist, Pollution Prevention Team Lead, Defense Logistics Agency*
3:15 PM	Considerations in Sustainable Sourcing and Procurement *Lorrie Vogel, General Manager, Nike's Considered Products, Nike Inc. (Teleconference)*
3:45 PM	Summary Remarks *Chair: Dara O'Rourke, University of California, Berkeley*
4:00 PM	MEETING ADJOURNS

Thursday, December 8, 2011

8:15 AM	Introduction and Previous Day Recap *Chair: Dara O'Rourke, University of California, Berkeley*

Session III: Analysis and Optimization with Sustainable Purchasing Tools
Moderator: Patrick Mallet, Founder, Credibility Director, ISEAL Alliance

8:30 AM	Current Landscape of Sustainable Procurement Tools *Chris O'Brien, Director of Sustainability, American University*

9:00 AM	Criteria for Best Practices with Sustainable Procurement Tools
	Josh Saunders, Sustainable Products Evangelist, Greencurement
9:30 AM	Opportunities and Emerging Requirements for Sustainable Procurement Tools
	Peter Graf, Chief Sustainability Officer, SAP
10:00 AM	Requirements for Procurement Processes Along the Supply Chain
	Edan Dionne, Director, Corporate Environmental Affairs, IBM Corporation
10:30 AM	BREAK
10:45 AM	A Framework for Greener Procurement: Connecting Life-cycle Attributes and Ecolabels to Bring Clarity to Institutional Buyers
	Jonathan Rifkin, Washington, DC and NASPO
	Yalmaz Siddiqui, Environmental Strategy Advisor, Office Depot
11:30 AM	LUNCH

Session IV: Moving Forward
Moderator: H. Scott Matthews, Professor, Carnegie Mellon University

12:15 PM	Components of a Proposed Sustainable Acquisition Tool
	Edward Rau, Chair, HHS Sustainability Innovations Working Group, National Institutes of Health
12:45 PM	A Roadmap for Achieving Sustainable Procurement
	Stephen Leeds, Senior Counselor to the Administrator, General Services Administration

APPENDIX A

1:15 PM   BREAKOUT GROUPS

What training and skills will be required to realize the full benefits of sustainable purchasing tools? What methods are there for broadening the use of sustainable purchasing tools to a more mainstream audience and encouraging a culture change among purchasers and suppliers?

Moderator: Chris O'Brien, Director of Sustainability, American University

Location: Conference Room A

What are major data gaps and data availability issues related to sustainable procurement? What are the different attributes, whether positive or negative, that characterize sustainable procurement and what should they be moving forward?

Moderator: Teresa Neal, Senior Purchasing Manager, Church & Dwight, Co., Inc.

Location: Board Room

2:15 PM   Breakout Group Report Back to Plenary
Moderator: H. Scott Matthews, Professor, Carnegie Mellon University

Location: Conference Room A

2:45 PM   Closing Remarks
Chair: Dara O'Rourke, University of California, Berkeley

3:00 PM   MEETING ADJOURNS

Appendix B

Registered Participants List

Sustainable Acquisition: Fostering Sustainability Considerations into Public and Private Sector Procurement Tools and Capabilities

20 F St. NW Conference Center
Washington, DC
December 7-8, 2011

Meadow Anderson
U.S. Environmental Protection Agency (US EPA)

David Asiello
U.S. Department of Defense

Leo Bonanni
Sourcemap

Dominic Brose
The National Academies

Erica Brown
Noblis

Yvonne Burgess
Climate Earth, Inc.

Scot Case
UL Environment

Kris Colby
Ariba

Brennan Conaway
U.S. General Services Administration (US GSA)

Mile Corrigan
Noblis

Richard Crespin
CROA

Alicia Culver
Responsible Purchasing Network

Jim Darr
US EPA

Edan Dionne
IBM

Janet Dobbs
US GSA

Beth Drake
US EPA

Jeff Eagan
U.S. Department of Energy (US DOE)

Jon Eisenberg
The National Academies

Holly Elwood
US EPA

Jonathan Estes
Facilities Solutions Group, LLC

Neal Etre
Industrial Economics

Shabnam Fardanesh
US DOE

Jim Fava
PE and Five Winds Strategic Consulting

Nancy Gillis
US GSA

Sheena Gilmore
International Trade Centre

Stephen Gordon
Old Dominion University

Peter Graf
SAP

Lara Greden
CA Technologies

Matt Haggerty
The National Academies

Deborah Hamilton
The Keystone Center

Brian Heath
U.S. Department of the Interior

Susan Hinton
U.S. National Institutes of Health (US NIH)

Rupert Hopkins
XSB

Ann-Marie Johnson
CSC

Adam Jones
US GSA

Emi Kameyama
The National Academies

Steven J. Kempf
US GSA

Alison Kinn Bennett
US EPA

Pat Koshel
The National Academies

Emily Lawrence
Booz Allen Hamilton

Eliza Lee
Veterans Affairs

Stephen Leeds
US GSA

C. Lindsay

APPENDIX B

Patrick Mallett
ISEAL Alliance

H. Scott Matthews
Carnegie Mellon University

Paul McRandle
NRDC

Katie Miller
US GSA

Lynette Millett
The National Academies

Karen Moran
Defense Logistics Agency

Marina Moses
The National Academies

Teresa Neal
Church & Dwight, Inc.

Ann Ngo
U.S. Department of Commerce

Chris O'Brien
American University

Syvera O'Pharrow
National Institute of Environmental Health Sciences

Anastasia O'Rourke
Big Room

Dara O'Rourke
University of California, Berkeley

Christopher Payne
Lawrence Berkeley National Laboratory

Edward Pfister
U.S. Department of Health and Human Services (US HHS)

Kristin Pierre
US EPA

Edward Rau
US NIH

Joshua Reese
Booz Allen Hamilton

Dylan Richmond
The National Academies

Jennifer Riddell
US EPA

Jonathan Rifkin
DC Office of Contracting and Procurement

Stephanie Rivera
US GSA

Anne Roberts-Smith
SAIC

Alec Rogers
Xerox

Kesa Russell
US HSS

Amanda Sahl
US DOE

Joshua Saunders
Greencurement

Rita Schenck
IERE

Kristen Sebasky
Industrial Economics

Nick Shufro
PwC

Yalmaz Siddiqui
Office Depot

Josh Silverman
US DOE

Amy Smith
World Wildlife Fund

Jan Stensland
Inside Matters

Joni Teter
US GSA

Peter Teuscher
BSD Consulting

Norma Tregurtha
iSEAL Alliance

Kathleen Turco
US GSA

Lorrie Vogel (teleconference)
Nike

Cindy Wasser
NACo

Andrew Weber
Lawrence Berkeley National Laboratory

Justin Yuen
FMYI

Eric Zoetmulder
SciQuest

Appendix C

Biographies

Fostering Sustainability Considerations into Public and Private Sector Procurement Tools and Capabilities

December 7-8, 2011
Washington, D.C.

BIOGRAPHICAL INFORMATION: PLANNING COMMITTEE, SPEAKERS AND STAFF

DARA O'ROURKE (Committee Chair) is associate professor in the Department of Environmental Science, Policy, and Management at the University of California, Berkeley. Previously, he was assistant professor in the Department of Urban Studies and Planning at the Massachusetts Institute of Technology. Dr. O'Rourke's research interests include the political economy and policy aspects global supply chains; governmental and nongovernmental strategies for monitoring and accountability over these production systems; and new models of public participation in environmental and labor policy regulation. He is currently leading a team of researchers focused on analyzing and improving the quality of information available to consumers on the social, environmental, and health impacts of products and companies; researching the impacts of this information on consumer behavior; and developing Web and mobile tools to deliver better information to consumers at their point-of-decision. He has recently applied this research to a social venture startup—GoodGuide.com—which provides information to consumers on the health, environmental and social performance of products and companies. Dr. O'Rourke previously served on the National Research Council's Committee Toward Improved International Labor Standards: Data, Monitoring, and Compliance. He completed his Ph.D. in the Energy and Resources program at the University of California at Berkeley.

ALISON KINN BENNETT is the senior advisor for product sustainability in EPA's Office of Pollution Prevention and Toxics. She co-founded and co-leads two influential, cross-media networks within EPA—the Green Building Workgroup and the Sustainable Products Network—which bring together policy and technical staff from around the agency in order to advance holistic, life-cycle based approaches to environmental and public health protection. Since 2001, Ms. Bennett has served in EPA's Environmentally Preferable Purchasing Program, focusing her efforts on coordinating EPA positions on standards and specifications for greener building products and construction services. Ms. Bennett is vice chair of ASTM International's Sustainability Committee (E60). Ms. Bennett earned a bachelor's degree in political science and geography from the University of California at Berkeley and a master's degree in urban and environmental planning from the University of Virginia's School of Architecture.

LEO BONANNI is the founder and CEO of Sourcemap.com, the crowd-sourced directory of product supply chains and carbon footprints. The open-source Web site offers tools for companies and individuals to share information about where things come from, what they are made of, and their social and environmental impact. Thousands of people have created sourcemaps of food, furniture, clothing, electronics, and more. Sourcemap's social network technology can also be used internally to help organizations gather supply chain information for traceability and risk management. Dr. Bonanni has a Ph.D. from the MIT Media Lab, an M.S. and a Master of Architecture from MIT, and a B.A. from Columbia University. He teaches sustainable product design at Parsons and at MIT.

DOMINIC A. BROSE (Staff) is a program officer for the Science and Technology for Sustainability Program (STS) at the National Academies. Prior to STS, Dr. Brose was with the Institute of Medicine (IOM) of the National Academies where he collaborated on science policy reports sponsored by the Department of Veteran Affairs (VA) addressing the potential for adverse health effects from exposure of select military personnel to environmental contaminants. Previously, he was an environmental scientist at ToxServices LLC, where he reviewed product formulations for EPA's Design for the Environment (DfE) program, which was a third-party service provided to EPA that evaluated product formulations against human health and environmental screening criteria. Dr. Brose received his M.S. and Ph.D. in environmental soil chemistry from the University of Maryland, and his B.S. in natural resources and environmental science from Purdue University.

BRENNAN CONAWAY has served as a contracting professional with General Services Administration's (GSA) Federal Acquisition Service (FAS) since 2004. Currently, Mr. Brennan works within GSA's Program Analysis Division, which is actively engaged in initiatives to green the agency's operations and supply chain. Brennan was awarded a B.B.A. from James Madison University and an M.B.A. from George Mason University.

ALICIA CULVER is the executive director of the Responsible Purchasing Network, an international network dedicated to advancing sustainable procurement policies and practices among government agencies and public institutions. Ms. Culver has over two decades of experience working in the sustainable procurement field. She got her start in 1994 evaluating the federal government's environmentally preferable purchasing efforts as Coordinator of the Government Purchasing Project based in Washington, D.C. She later served as deputy director of the New Jersey Office of Sustainability and, in 2004, founded the Green Purchasing Institute. Ms. Culver is currently serving as an advisor to the UN Environment Program and the World Bank, identifying the best practices for the procurement of energy-efficient products around the world. She also chairs San Francisco's Sweatfree Procurement Advisory Group. She has co-authored many publications on sustainable procurement, including *Cleaning for Health: Products and Practices for a Safer Indoor Environment*, *RPN's Responsible Purchasing Guide to LED Lighting*, and *Buying Smart: Experiences of Municipal Green Purchasing Pioneers*.

EDAN DIONNE is director of corporate environmental affairs at IBM Corporation. IBM's Corporate Environmental Affairs staff sets the company's global strategy for and oversees IBM's programs and performance worldwide in environment, energy and climate, product and process environmental stewardship. She joined IBM in 1983 as environmental engineer and became part of IBM's corporate environmental affairs staff in 1990. Prior to assuming her current position, her experience included managing a wide range of environmental and climate protection programs, partnership with others in industry, the USEPA and nongovernmental organizations. She assumed her present position in 2002. Ms. Dionne has a master of science degree in chemical engineering.

JON EISENBERG (Staff) is director of the Computer Science and Telecommunications Board of the National Academies. At CSTB, he has also been study director for more than a dozen major studies, including a series of reports exploring Internet and broadband policy and networking and communications technologies. From 1995 to 1997 he was a AAAS Science,

Engineering, and Diplomacy Fellow at the U.S. Agency for International Development, where he worked on technology transfer and information and telecommunications policy issues. Dr. Eisenberg received his Ph.D. in physics from the University of Washington in 1996 and a B.S. in physics with honors from the University of Massachusetts at Amherst in 1988.

NANCY GILLIS directs the GSA Federal Supply Chain Emissions Program Management Office (PMO), which is tasked to create and promote a more sustainable federal supply chain. She chairs the Section 13 Interagency Working Group, addressing the technical and policy recommendations outlined in the EO 13514 Vendor and Contractor Emissions report, and manages the GreenGov Supply Chain Partnership, a public/private collaboration seeking to reduce environmental impacts throughout the supply chain. Ms. Gillis has spent the majority of her career in the area of sustainability and has worked internationally on biodiversity, supply chain, economic development, and technology innovation projects. Ms. Gillis received her graduate degree from Georgetown University and is a proud alumnus of the Santa Fe Institute Complexity program.

STEPHEN GORDON directs the Graduate Certificate Program in Public Procurement and Contract Management in the Department of Urban Studies and Public Administration at Old Dominion University in Norfolk, Virginia. Experienced as a manager, leader, thinker, and innovator in the not-for-profit, for-profit, and public sectors, Dr. Gordon teaches graduate classes in public procurement and contract management and public policy. Dr. Gordon's primary areas of research interest include performance-based contracting, sustainable procurement, out-sourcing and in-sourcing of public services, and relationships in an intergovernmental and multi-sector environment. Throughout his career, Dr. Gordon has contributed to the advancement of the practice of governmental procurement, especially at the state and local levels. He co-chairs the Steering Group of the Sustainable Procurement Initiative, a group of professionals with differing but complementary expertise and perspectives that came together in 2010 to jointly focus on cost-effectively organizing governmental procurement in the United States to promote sustainability. The Sustainable Procurement Initiative is developing an organized, market-driven, non-regulatory strategy involving performance-based acquisition, contract incentives, large scale cooperative contracts, and information sharing and networking to dramatically increase sustainable purchases and contracts.

PETER GRAF (Committee Member) is chief sustainability officer and executive vice president of sustainability solutions with SAP, where he is responsible for developing sustainable solutions that best serve the needs of SAP's global customers, while also driving sustainable operations within SAP. At SAP, Dr. Graf has held various management roles. Previously, he was the executive vice president of solution marketing at SAP. In this role, he was responsible for shaping the company's industry solution, application, and platform strategy. Dr. Graf holds a master's degree in computer science and economics as well as a Ph.D. in artificial intelligence.

STEVEN J. KEMPF was appointed commissioner for the U.S. General Services Administration's Federal Acquisition Service (FAS), effective July 10, 2010. In this capacity, he sets strategic direction and oversees the delivery of over $50 billion of best-value products, services, and solutions to federal customers, allowing them to effectively and efficiently achieve their missions. He also held this position in an acting capacity from April through June 2010, and was the deputy commissioner prior to that. Mr. Kempf has held multiple leadership positions throughout FAS and its predecessor organizations. He served as the acting FAS deputy commissioner from October 2008 through January 2009. In February 2008, he was named assistant commissioner for the FAS Office of Acquisition Management, where he was responsible for overall acquisition policy planning and coordination. Mr. Kempf holds a B.A. in history from Marquette University in Milwaukee, Wisconsin, and a law degree and an M.B.A. from the George Washington University in Washington, D.C.

STEPHEN LEEDS[1] was appointed as senior counselor to the administrator for the U.S. General Services Administration (GSA) on August 10, 2009. As senior counselor, Mr. Leeds advises and assists the administrator on a variety of GSA's enterprise initiatives. He also coordinates the activities of the regional administrators. Mr. Leeds is the senior accountable official for recovery, placing him in charge of the agency's efforts to implement the American Recovery and Reinvestment Act. He is also the agency senior sustainability officer, leading GSA's efforts under Executive Orders 13423 and 13514 to fulfill GSA's responsibilities and opportunities under those EOs as well as assisting GSA's federal agency customers with solutions to help them integrate sustainability throughout their agencies and achieve their sustainability goals. Mr. Leeds graduated from Michigan State University with a B.A. in business and received a J.D. from the University of Michigan.

---

[1] Senior Counselor to the Administrator as of December 7, 2011.

PATRICK MALLET (Committee Member) is founder and credibility director with ISEAL Alliance, the global association for social and environmental standards. At ISEAL Alliance, Mr. Mallet is responsible for managing the development of consensus-based codes of good practice for the effective operation of voluntary standards systems. In 2004, he led the development of the ISEAL Code of Good Practice for Setting Social and Environmental Standards which has since become the normative reference point for credible standard-setting practices. Prior to founding the ISEAL Alliance in 2000, Mr. Mallet managed an international program in certification and marketing of non-timber forest products and was lead author on the multi-stakeholder Conservation Principles for Coffee Production. He is past board chair of the Certified Organic Associations of British Columbia. He earned his degrees at Dalhousie and McGill Universities in eastern Canada.

H. SCOTT MATTHEWS (Committee Member) is the research director of the Green Design Institute and professor in the Department of Civil and Environmental Engineering and the Department of Engineering and Public Policy at Carnegie Mellon University. His work includes valuing the socioeconomic implications of environmental systems and infrastructure and industrial ecology. He focuses on using the Internet to facilitate environmental life-cycle assessment of products and processes, estimating and tracking carbon emissions across the supply chain, and the sustainability of product systems and infrastructure as connected to public policy. Dr. Matthews previously served on the National Research Council Committee on Health, Environmental, and Other External Costs and Benefits of Energy Production and Consumption. He holds a Ph.D. in economics from Carnegie Mellon University.

LYNETTE I. MILLETT (Staff) is associate director at the Computer Science and Telecommunications Board (CSTB), National Research Council of the National Academies. She currently directs several CSTB projects, including an investigation into foundational science in cybersecurity and an examination of opportunities for computing research to help meet sustainability challenges. She served as the study director for the CSTB reports *Strategies and Priorities for Information Technology at the Centers for Medicare and Medicaid Services and Social Security Administration Electronic Service Provision: A Strategic Assessment*. Ms. Millett's portfolio includes significant portions of CSTB's recent work on software, identity systems, and privacy. She directed, among other projects, those that produced *Software for Dependable Systems: Sufficient Evidence?*, an exploration of fundamental approaches to developing dependable mission-critical systems; *Biometric Recognition: Challenges and Opportunities*, a comprehensive assessment of biometric technology; *Who Goes There? Authentication Through the Lens of Privacy,* a discussion of authentication technologies and their privacy implications; and *IDs—Not That Easy: Questions About Nationwide Identity Systems,* a post-9/11 analysis of the challenges presented by large-scale identity systems. She has an M.Sc. in computer science from Cornell University and a B.A. in mathematics and computer science from Colby College.

KAREN MORAN is currently the team lead for Pollution Prevention at Headquarters Defense Logistics Agency. She has over 30 years of experience in Environmental, Safety, and Occupational Health management with the Department of Defense. Her assignments have included installations, Headquarters for Major Commands & Components, and the Pentagon. She holds B.S. and M.S. degrees from the College of William and Mary in Virginia and the University of Southern California.

MARINA S. MOSES (Staff) serves as the director for the Science and Technology for Sustainability Program (STS) in the Division of Policy and Global Affairs of the National Academies. In this capacity, she also serves as the director of the Roundtable on Science and Technology for Sustainability. Under her leadership, the STS program issued the consensus report Sustainability and the U.S. Environmental Protection Agency and has recently undertaken the multi-sponsored study Sustainability Linkages in the Federal Government. Prior to joining the National Academies, Dr. Moses served on the faculty of the George Washington University School of Public Health and Health Services in the Department of Environmental and Occupational Health, where she also directed the Doctoral Program and was the practicum coordinator for the graduate program. Dr. Moses was the recipient of the 2005 Pfizer Scholar in Public Health Award and has worked in emergency preparedness and communication with communities on public health issues. Previously, Dr. Moses held senior scientific positions in the Environmental Management Division of the U.S. Department of Energy, where she worked on the development of a qualitative framework to assess hazardous and nuclear waste risks, and served as the senior health risk assessor in the New York City office of the U.S. Environmental Protection Agency's Superfund Program. Dr. Moses received her Bachelor of Arts (Chemistry) and her Master of Science (Environmental Health Sciences) degrees from Case Western Reserve University. She received her doctorate of public health (Environmental Health Sciences) from Columbia University School of Public Health.

TERESA NEAL (Committee Member) is senior purchasing manager at Church & Dwight, Co. Inc. Formerly, Dr. Neal was the product marketing director for SciQuest's Commercial sector, where she was responsible for product strategy and driving growth in the emerging commercial markets and served as SciQuest's thought leader for Green Procurement. Before joining SciQuest, she was a senior marketing manager for North America and Latin America at Novozymes, where she positioned enzymes as Green ingredients in the household care industry. She has an M.B.A. from the University of North Carolina, Greensboro, and a Ph.D. in bioinorganic chemistry from the University of Notre Dame.

CHRIS O'BRIEN joined American University in 2009 as the university's first director of sustainability. He is responsible for sustainability policy, planning, outreach, and implementation. His work includes leading the university's commitment to achieving climate-neutrality by the year 2020, as well as building and operation green buildings, eliminating waste sent to landfill, and greening the university's procurement. Previously, he directed the Responsible Purchasing Network at the Center for a New American Dream, and earlier served as Managing Director of the Green Business Network and the Fair Trade Federation. He is treasurer of the Fair Trade Resource Network and co-owns the Seven Bridges Organic Brewing Supply Cooperative. He serves on the Electronic Products Environmental Assessment Tool (EPEAT) Advisory Board, the Green Advantage Board, and the Association for the Advancement of Sustainability in Higher Education's STARS Steering Committee. In 2006, Chris authored the award-winning book *Fermenting Revolution: How to Drink Beer and Save the World* (New Society Publishers). He has a bachelor's degree in liberal arts from Penn State University and a master's degree in science and technology studies from Rensselaer Polytechnic Institute.

ANASTASIA O'ROURKE is a co-founder of Big Room Inc.—the creators of Ecolabel Index (www.ecolabelindex.com), and a proposal for a new top level domain (*www.doteco.org*). Big Room also provides advisory services to clients such as the GSA, UNEP, and FedEx Corporation. She co-chairs the Green Products Roundtable, a multi-stakeholder group of corporations, nonprofits, certifiers, and government who are working together to find ways to help bring clarity to the green marketplace. Dr. O'Rourke is an expert in designing information platforms that facilitate the growth of green markets and in assessing sustainability of companies, from her recent work on green products and certification systems, to prior work on cleantech venture investments, socially responsible investment metrics, and corporate and governmental environmental reporting. She has a Ph.D. from Yale University and an M.Sc. from Lund University, both in environmental management. She has lived and worked in Australia, Sweden, France, and the United States, where she currently resides.

EDWARD RAU serves as chair of the U.S. Department of Health and Human Services Sustainability Innovations Working Group at the National Institutes of Health (NIH). He has a 30-year tour of active duty as an Environmental Health Officer in the U.S. Public Health Service Commissioned Corps, continuing in civilian service with the NIH. He has served initially as chemical waste manager, advanced to chief of hazardous and solid waste management and special assistant to the director, Division Environmental Protection, NIH. Captain Rau has a bachelor's degree in biology and an M.S. in environmental and occupational health sciences from California State University at Northridge, and a graduate certificate in hazardous materials management from the University of California, Davis.

DYLAN RICHMOND (Staff) is a research assistant for the Science and Technology for Sustainability Program (STS) at the National Academies. Before joining the Academies in the fall of 2010, he attended Georgetown University and graduated with a B.A. in economics in May 2010. While at Georgetown, Mr. Richmond was an editor for *The Georgetown Voice*.

JONATHAN RIFKIN is a special assistant to the director for the District of Columbia Office of Contracting and Procurement (OCP). His work in Green Procurement began approximately four years ago when he was given the opportunity to work for OCP. He has spent the last year developing a green purchasing program within the District, which continues to mature. He also serves on the District's "DC Green Team," which is the District's unified effort to enact sustainable practices across the government. Most recently, he was appointed to serve as liaison to the District's Mayor's Sustainability Initiative, which will establish a comprehensive sustainable strategy for the District for the approaching decades. In addition, Jonathan serves on the National Association of State Procurement Officers (NASPO) Green Purchasing Work Group, which works to identify and share Green Purchasing best practices with its membership; and the Keystone Group's Green Products Roundtable, which is a multi-stakeholder group dedicated to bringing clarity to the green marketplace and endorsing practices that simplify and maximize green purchasing efforts for Institutional Purchasers. He is fascinated by the complexities of the green marketplace and takes enormous satisfaction from the positive impact that environmentally preferable purchasing can have on his local and global world.

JOSH SAUNDERS is a nationally recognized expert on product-level sustainability and environmental certifications. Joshua is the co-founder and CEO of Greencurement, a company that provides environmental information for institutional purchasing. At Greencurement, Joshua works with leading organizations to identify green products and make smarter purchasing decisions. Prior to Greencurement, Joshua was the senior director of business development for GoodGuide, where he was responsible for partnerships, alliances, and commercial sales. Joshua has also worked closely with a number of environmental standard setting, ecolabeling, and certification programs across a wide range of industries from electronics to building products. Joshua was a co-founder of UL Environment, a subsidiary of Underwriters Laboratories, where he led global operations and oversaw the environmental claims verification and sustainable products certification programs. He has been heavily involved in many sustainability initiatives, including participating in past roundtables with the National Academies and the Keystone Center as well as being a contributor to NPR's Marketplace Report. Mr. Saunders holds a bachelor's degree in electrical and computer engineering and an M.B.A. in finance and entrepreneurship.

YALMAZ SIDDIQUI is Office Depot's senior director of environmental strategy. He is responsible for setting strategy, defining metrics, and driving integration of a wide range of environmental programs into the global organization. Under his leadership, Newsweek magazine ranked Office Depot in 2011 as the #1 greenest large retailer and the #8 greenest large U.S. company overall. Mr. Siddiqui is on the Board of Advisors of EPEAT; is the co-chair of the Green Products Roundtable, a multi-stakeholder group working on defining 'green products' for institutional purchasers; and was the only private sector recipient of the Responsible Purchasing Network's 2009 Responsible Purchaser of the Year Award. He holds a master of philosophy in environment & development from the University of Cambridge, where his thesis focused on industrial ecology, life-cycle analysis, and bio-mimicry. He also has a bachelor of commerce degree from McGill University in Canada.

JOSH SILVERMAN is the director of the Office of Sustainability Support at the U.S. Department of Energy. His office provides training, technical assistance, and corporate reporting and analysis support to help DOE achieve its sustainability goals. He engages in a broad spectrum of activities within DOE, including sustainable acquisition, sustainable buildings, green electronics, and toxic chemical reduction. He recently received an Achievement Award from the Secretary of Energy for helping DOE reduce fugitive emissions of potent greenhouse gases. Dr. Silverman joined DOE in 2000, after completing a dissertation at Carnegie Mellon University on environment, safety, and health practices in nuclear weapons production.

LORRIE VOGEL, general manager of Considered Design, Nike's sustainable design team, seeks to intertwine design innovation and conservation. Through Considered Design, Ms. Vogel's focus is to design out waste, chemicals and energy, and design in new materials and approaches. Prior to becoming the general manager, she was the innovation director for Nike Footwear, Apparel, and Equipment. Her innovative thinking has led to several new product technologies and patents. Her passion for design, innovation and sustainability together brought her to the role as a leader in bringing sustainability to all facets of Nike products. With an industrial design degree from Syracuse University, Ms. Vogel has become a leading expert in design innovation. She has worked for many companies in product design and research and development for Texas Instruments and S.G. Hauser, a top design consulting firm in Los Angeles. In early 2007 Ms. Vogel was named to Fast Company's Fast 50, the magazine's annual compilation of innovative companies and the "incubators of genius." Her team's innovative work around sustainable design has been recognized in several publications: *Strategies for Sustainable Design* (Adam Worbach), *Green to Gold* (Dan Esty), and as a M.I.T. Case Study.

Appendix D

Examples of Federal Agency Procurement Systems and Green Purchasing Programs

Procurement System	Description
	Department of Agriculture
Integrated Acquisition System (IAS)[1]	IAS is a Web-based eProcurement system designed to streamline and automate contract management and acquisition processes throughout USDA. USDA Advantage, part of IAS e-alliance initiative, was formed as a partnership between USDA and GSA to better leverage USDA spending power to obtain better pricing, ordering methods, and delivery terms for commonly acquired items and services.
USDA Purchase Card Management System (PCMS)[2]	PCMS is a customized, Web-based online relational database used to manage purchases made with government-wide commercial purchase cards, convenience checks, or fleet cards issued by an authorized GSA contractor.
USDA BioPreferred program[3]	The USDA BioPreferred program promotes the increased purchase and use of biobased products, which is expected to reduce petroleum consumption, increase the use of renewable resources, better manage the carbon cycle, and may contribute to reducing adverse environmental and health impacts.

Procurement System	Description
	Department of Defense
DOD EMALL[4]	DOD EMALL is an Internet-based mall that allows military and other authorized government customers to procure items from government and commercial sources. DOD EMALL is a Department of Defense program operated by the Defense Logistics Information Service.
DoD Standard Procurement System (SPS)[5]	The SPS program was created to standardize the procurement process throughout DOD by developing, testing, and deploying a suite of software products for use by contracting professionals in the Army, Navy, Air Force, Marines, and other defense agencies.
Defense Acquisition System[6]	The Defense Acquisition System is the management process that guides all DOD acquisition programs, provides the policies and principles, and establishes the management framework that implements these policies and principles.
	Department of Energy
Federal Energy Management Program[7]	The DOE Federal Energy Management Program (FEMP) supports federal agencies in identifying energy- and water-efficient products that meet federal acquisition requirements (FAR), conserve energy, save taxpayer dollars, and reduce environmental impacts. This is achieved through technical assistance, guidance, and efficiency requirements for energy-efficient, water-efficient, and low standby power products.
ENERGY STAR (Joint program with EPA)[8]	ENERGY STAR, a joint DOE and EPA program, delivers technical information and tools that organizations and consumers use to select energy-efficient solutions and best management practices. This program encourages and assists governments, schools, and businesses in procuring ENERGY STAR qualified and FEMP designated products.
Strategic Integrated Procurement Enterprise System (STRIPES)[9]	The Strategic Integrated Procurement Enterprise System (STRIPES) is the procurement and contracts management component of the DOE's Integrated Management Navigation System program and encompasses both acquisition and financial assistance. The STRIPES project reduced the number of procurement-related systems across the department.

APPENDIX D   57

Procurement System	Description
	Environmental Protection Agency
EPA's WaterSense Program[10]	WaterSense is a partnership program by the U.S. Environmental Protection Agency that aims to protect the nation's water supply by offering people a simple way to use less water with water-efficient products, new homes, and services. The goal is to provide consumers with a label for products and an information resource to help people use water more efficiently; encourage innovation in manufacturing; and decrease water use and reduce strain on water resources and infrastructure. Products and services with the WaterSense label have been certified to be at least 20 percent more efficient without sacrificing performance.
Comprehensive Procurement Guideline (CPG)[11]	The CPG program is a continuing effort to promote the use of materials recovered from solid waste. The CPG program is authorized by Congress under section 6002 of the Resource Conservation and Recovery Act, and EPA is required to designate products that are or can be made with recovered materials and to recommend practices for buying these products. Procuring agencies are then required to purchase designated products with the highest recovered material content level practicable. Recovered Materials Advisory Notices provide purchasing guidance and recommend recovered and postconsumer material content levels for designated items.
Environmentally Preferable Products (EPP)[12]	The EPP is a database to facilitate the purchase of products and services with reduced environmental impacts by providing information on mandatory purchase programs contained in FAR Part 23 (Energy Star, CPG, and Biopreferred Products).
Significant New Alternatives Policy (SNAP)[13]	SNAP is EPA's program to evaluate and regulate substitutes for the ozone-depleting chemicals that are being phased out under the stratospheric ozone protection provisions of the Clean Air Act (CAA).
	General Services Administration
GSA Advantage[14]	The GSA Advantage is a one-stop online resource to help agencies better serve the public by meeting their needs for products and services, and to simplify access to information.

Procurement System	Description
E-Buy[15]	GSA's E-Buy is an online procurement tool that allows buyers to issue requests for quotations for products and services offered on GSA multiple award schedules. E-Buy is designed to be complementary to GSA Advantage and is available for use by federal buyers.
GSA Global Supply[16]	GSA Global Supply's online ordering site that has access to approximately 400,000 tools, office supplies, computers, and other supplies.
Green Products Compilation[17]	Green Products Compilation (GPC) is designed to facilitate the procurement of green/sustainable products and services. The products listed are those for which the EPA, DOE, and USDA issued designations or guidance for environmental or energy attributes.
National Aeronautics and Space Administration	
NASA Acquisition Internet Service (NAIS)[18]	NAIS is an online service that grants industry immediate access to current acquisition information. NAIS is a feeder system for federal E-Gov systems like the Federal Business Opportunities (FBO). NAIS provides industry links to reference information such as regulations, clauses, provisions, handbooks, and guidance documents. NAIS also provides industry a central location to find each NASA field center's procurement home page. Additionally, NAIS provides NASA procurement professionals with a suite of procurement tools for internal use.
Recycling and Sustainable Acquisition (RSA)[19]	NASA's Lead Center for Recycling and Sustainable Acquisition (RSA) provides technical resources and program support for NASA Headquarters Environmental Management Division.
National Institutes of Health	
NIH Green Purchasing Program[20]	The NIH Green Purchasing Program involves the selection and acquisition of products and services that most effectively minimize negative environmental impacts over their life cycle of manufacturing, transportation, use and recycling or disposal. The Green Purchasing Program helps NIH improve safety and health of patients, workers and the public; reduce pollution; conserve natural resources; stimulate new markets for recycled materials, improve environmental stewardship; provide potential cost savings; and comply with environmental laws and regulations.

NOTES

[1] *www.afm.ars.usda.gov/initiatives/ias/*

[2] *www.dm.usda.gov/procurement/card/pcms.htm*

[3] *www.biopreferred.gov/?SMSESSION=NO*

APPENDIX D

[4] *www.landandmaritime.dla.mil/programs/emall/*
[5] *http://sps.caci.com/*
[6] *https://dap.dau.mil/aphome/das/Pages/Default.aspx*
[7] *www1.eere.energy.gov/femp/*
[8] *www.energystar.gov/*
[9] *http://energy.gov/management/downloads/pia-imanage-strategic-integrated-procurement-enterprise-system-stripes*
[10] *www.epa.gov/watersense/*
[11] *www.epa.gov/epawaste/conserve/tools/cpg/about.htm*
[12] *www.epa.gov/opptintr/epp/index.htm*
[13] *www.epa.gov/ozone/snap/*
[14] *www.gsaadvantage.gov*
[15] *www.ebuy.gsa.gov*
[16] *www.gsaglobalsupply.gsa.gov*
[17] *www.gsa.gov/portal/content/198257*
[18] *http://prod.nais.nasa.gov/cgi-bin/nais/index.cgi*
[19] *http://nasarecycles.nasa.gov/*
[20] *http://nems.nih.gov/programs/Pages/purchasing.aspx*

Appendix E

The Federal Life-Cycle Assessment (LCA) Digital Commons[1]

Life-cycle assessment is the assessment of environmental, economical, and social impact of a product throughout its lifespan, and is increasingly important in product design, procurement decisions, and policy making in the public and private sectors. The USDA, EPA, DOE, and other federal agencies have generated and are continuing to produce a critical mass of data available to support LCA activities. However, there is currently a lack of reliable, comprehensive, and open access data to support life-cycle assessments (LCA) for public and private decision making.

The primary purpose of the federal LCA Digital Commons is to fully leverage federal investments in sustainability science and technology to maximize benefits by coordinating LCA data collection, organization, management, dissemination, and preservation. This project adopts an open source, open standard, and open access approach for system development. Stated deliverables from the federal LCA Digital Commons include:

- Peer-reviewed LCA data, which ensures validity and reliability for analyses by providing reliable and comprehensive data made available through open database access
- Data preservation
- Standards and protocols, which are open, practical, and matured standards to represent and exchange LCA data and models

---

[1] *www.lcacommons.gov/*

- Nomenclatures, which provide a shared vocabulary and naming convention to represent LCA data and model LCA methods and practices
- Tools to collect, organize, validate, analyze, and visualize LCA data and models, and services to facilitate activities in the LCA community